S0-CNI-983

Oil, Gas, and Coal

Jacqueline Dineen

RAINTREE
STECK-VAUGHN
PUBLISHERS
The Steck-Vaughn Company

Austin, Texas

Editor: Claire Llewellyn
Series Editor: Pippa Pollard
Science Editor: Kim Merlino
Design: Caroline Ginesi
Project Manager: Julie Klaus
Electronic Production: Scott Melcer
Artwork: Ian Thompson
Cover Artwork: Ian Thompson
Picture Research: Brooks Krikler Research, Juliet Duff

Library of Congress Cataloging-in-Publication Data
Dineen, Jacqueline.
Oil, gas, and coal /Jacqueline Dineen.
p. cm. — (What about?)
Includes index.
ISBN 0-8114-5535-1
1. Petroleum engineering — Juvenile literature. 2. Coal mines and mining — Juvenile literature. [1. Petroleum. 2. Gas 3. Coal mines and mining.] I. Title. II. Series.
TN870.3.D56 1995
553.2—dc20 94-35144
CIP
AC

Printed and bound in the United States by Lake Book, Melrose Park, IL

1 2 3 4 5 6 7 8 9 0 LB 98 97 96 95

Contents

What Are Oil, Gas, and Coal? 3
How Were Oil and Gas Formed? 4
Finding Oil and Gas 6
Drilling an Oil Well 8
Finding Oil Under the Ocean Floor 10
At the Oil Refinery 12
Finding and Using Gas 14
How Was Coal Formed? 16
Mining Coal 18
An Underground Factory 20
Cutting the Coal 22
Bringing the Coal Out 24
Surface Mining 26
Using Coal 28
Things to Do 30
Glossary 31
Index 32

What Are Oil, Gas, and Coal?

Oil, gas, and coal are types of fuel. They were formed from plants, animals, and other living things that died many millions of years ago. The remains of these living things are called **fossils**. That is why oil, coal, and gas are known as fossil fuels. They are usually buried deep in the earth under layers of rock.

▽ Fuel can be burned to release energy. The energy in coal is changed into heat when it burns.

How Were Oil and Gas Formed?

Oil and gas form over millions of years from tiny living things in the ocean. When these died, they sank to the bottom and were buried in the sand. Rivers carried more rock and sand to the ocean. Layers of rock and sand were pressed down so tightly that they formed **sedimentary rock**. The living things **decayed** in the rocks and changed into oil and gas.

▽ An oil rig is set up to drill for oil under the ocean floor.

1. Oil and gas are found in places that were once covered by ancient oceans and lakes.

2. Living things died and sank to the bottom of the ocean.

3. They were slowly buried under layers of sand, which pressed down hard on the remains of these creatures.

4. Over millions of years the sand turned into rock. Heat and pressure turned the remains into oil and gas.

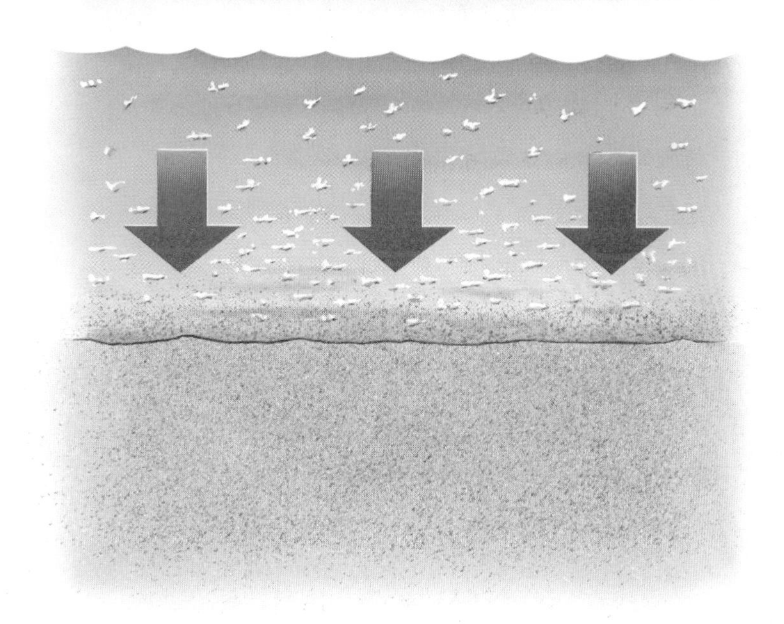

Finding Oil and Gas

Over millions of years, land was pushed up out of the ocean by movements in the Earth. This explains why oil and gas are sometimes found under dry land. They are usually buried so deep that scientists have to work hard to find them.

Geologists study the rocks on the surface and make **surveys** to find out more about the layers underneath.

▽ Geologists drill into the Earth and take out a sample that shows the rock layers.

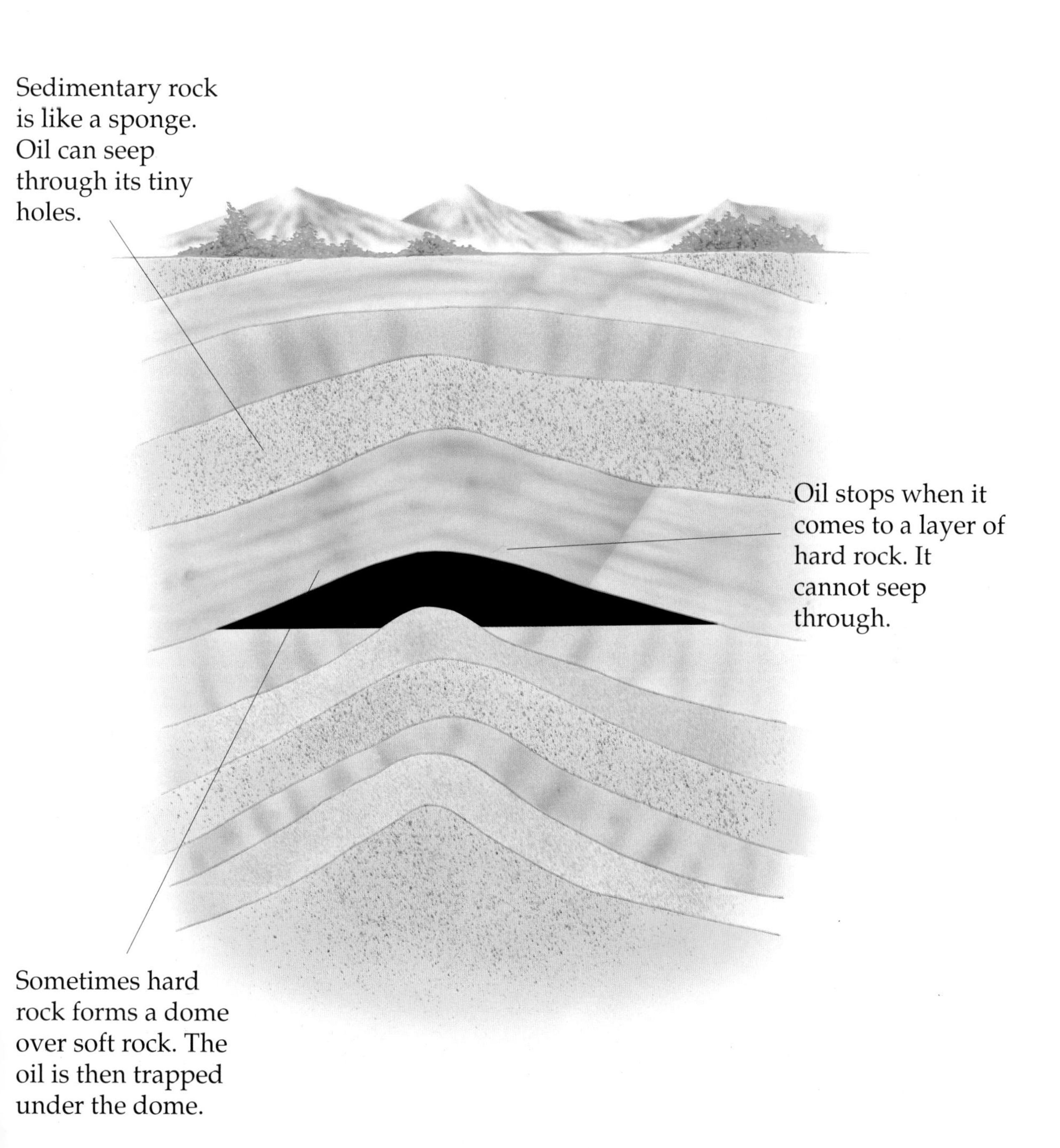
Sedimentary rock is like a sponge. Oil can seep through its tiny holes.
Oil stops when it comes to a layer of hard rock. It cannot seep through.
Sometimes hard rock forms a dome over soft rock. The oil is then trapped under the dome.

Drilling an Oil Well

When geologists think they have found oil, a well is drilled to see if they are right. Under the ground, oil is forced upward by **pressure**. When the drill reaches the oil, there may be enough pressure to send it gushing to the surface. If not, it has to be pumped up. The oil is then sent by pipeline to a **refinery** and is turned into products such as gasoline.

▷ Some oil may be piped to a port and loaded onto oil tankers for shipping overseas.

▽ As the drill goes deeper, more lengths of pipe are added.

▷ A rotary drill is made up of lengths of pipe. The cutting tool is called the bit.

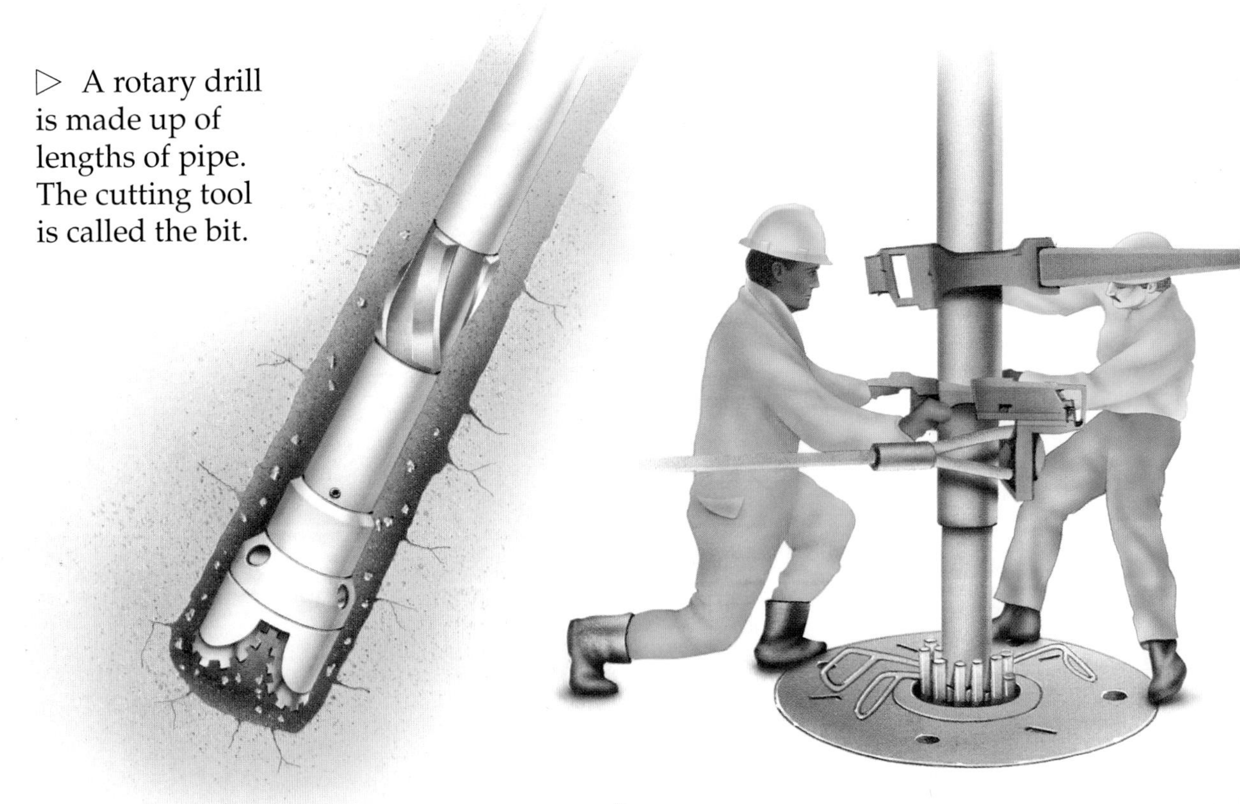

Finding Oil Under the Ocean Floor

Oil has been found under the seafloor. Then a platform is put in place so that production can begin. Oil workers live on a separate platform for a couple of months at a time. The work never stops, so the crews have to work in shifts. As one team finishes, another one begins. A pipeline is laid along the seafloor so that the oil can be piped to the shore.

▽ Before the oil platform is set up, the oil company runs tests. They use an exploration rig to see if there is oil.

▷ An oil rig drills a well down to oil trapped below the ocean floor. The production platform brings the oil to the surface.

It is a hard life working on an offshore oil rig. Workers relax on a separate platform where there are cabins, kitchens, a restaurant, and a movie theater.

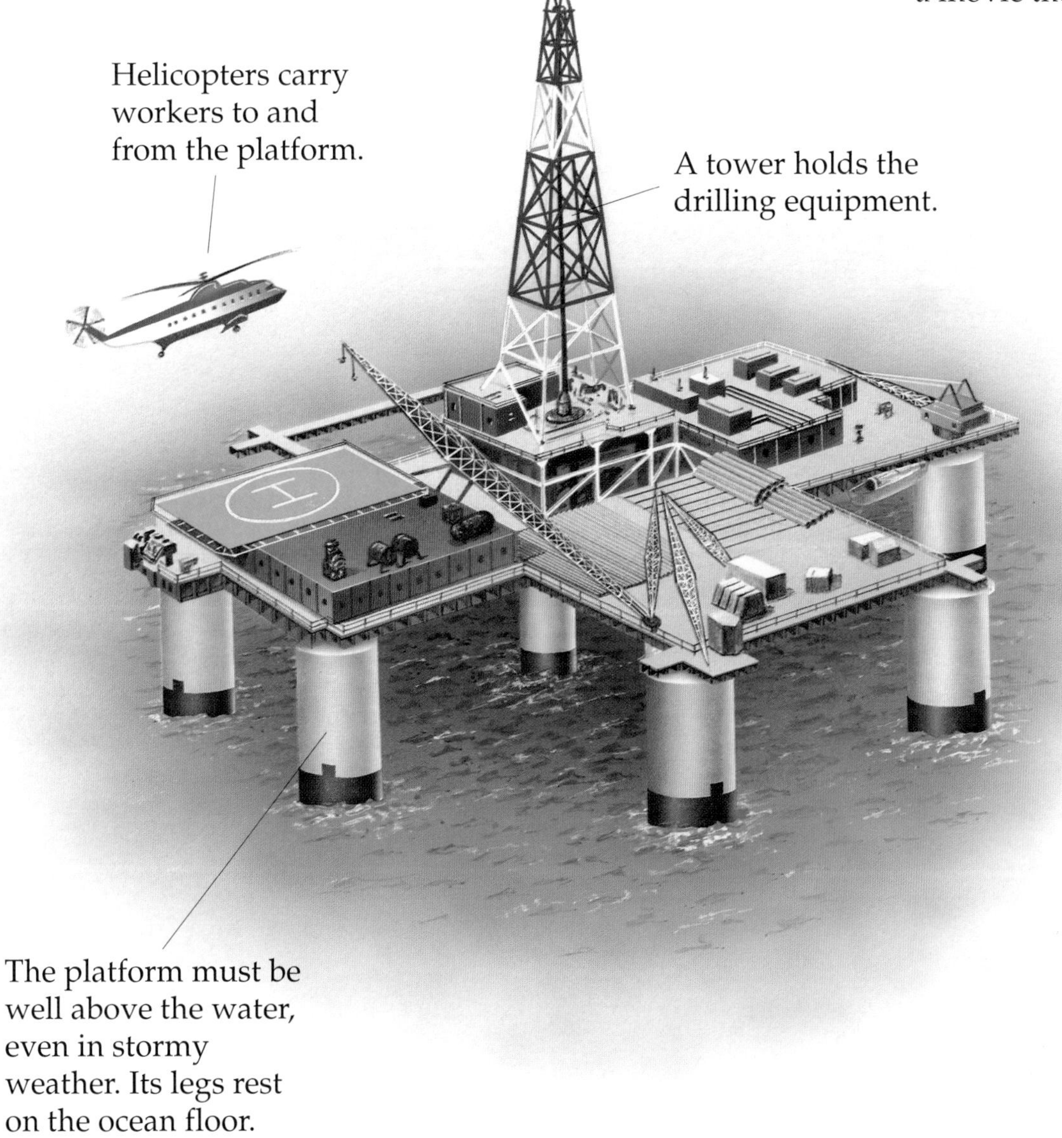

Helicopters carry workers to and from the platform.

A tower holds the drilling equipment.

The platform must be well above the water, even in stormy weather. Its legs rest on the ocean floor.

At the Oil Refinery

▷ Machinery is oiled to keep the parts moving smoothly.

The oil that comes out of the ground is called crude oil. It is dirty and contains dangerous gases. Before it can be used, it has to be separated into different parts, and then cleaned and treated. This happens at the refinery. The oil is made into products such as gasoline, diesel oil, and kerosene. The oil products are taken to customers in tank trucks or rail tankers.

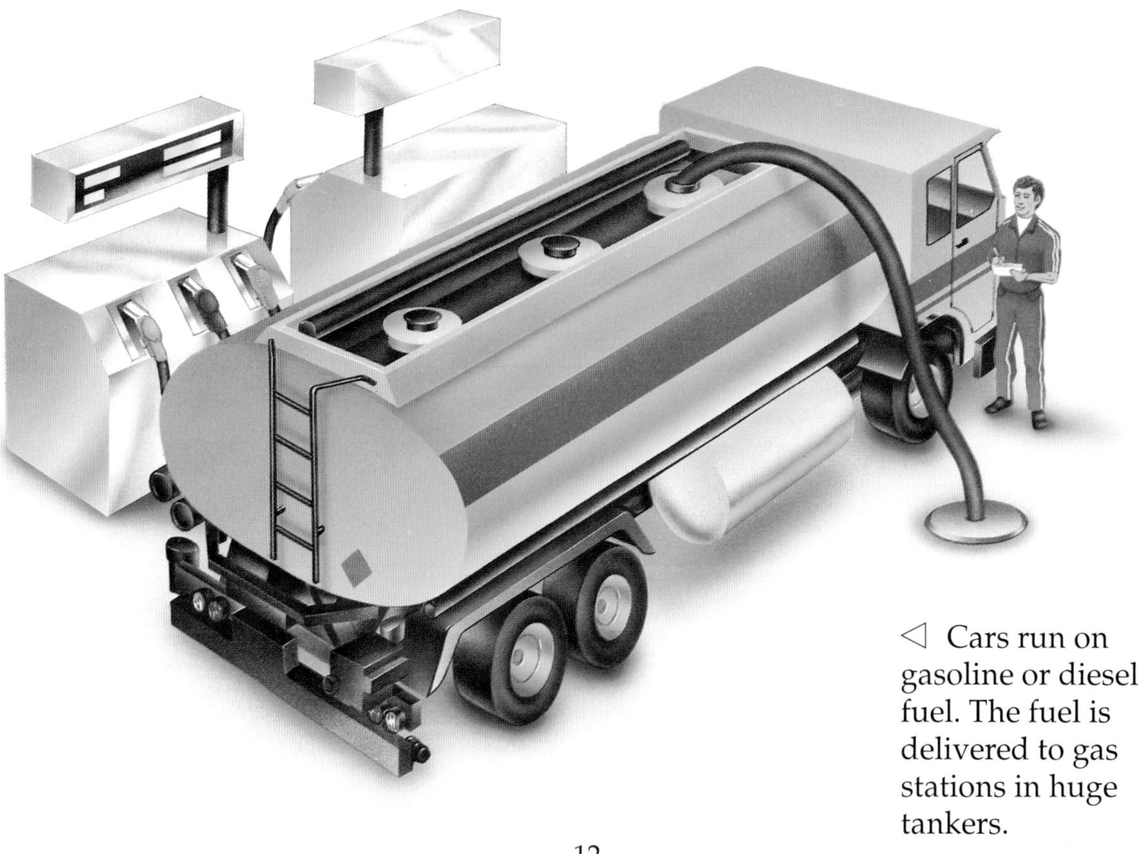

◁ Cars run on gasoline or diesel fuel. The fuel is delivered to gas stations in huge tankers.

▷ Another product made from oil is asphalt. This is used in road building.

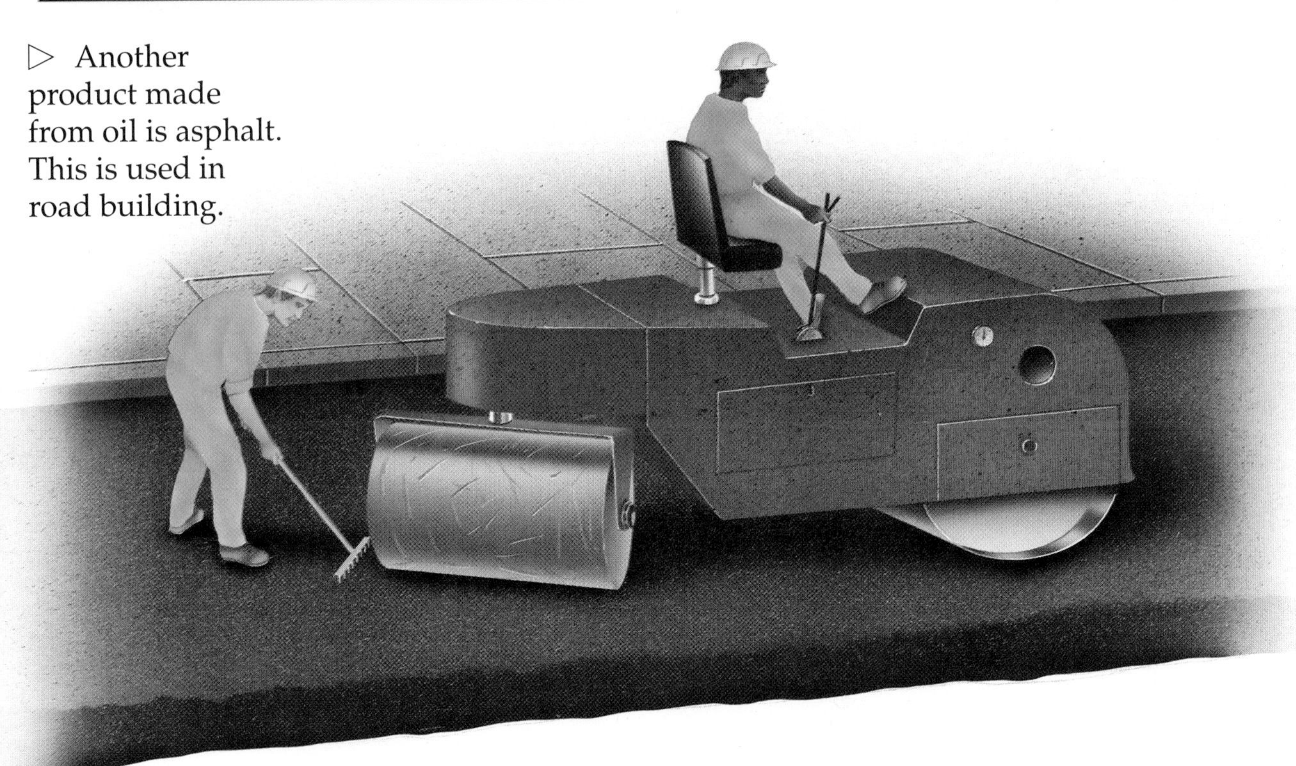

Finding and Using Gas

Natural gas is often found in the same places as oil and is brought to the surface at the same time. It is separated from the oil and carried by pipeline to a natural gas processing plant. Here it is cleaned, and the unwanted parts of the gas are removed so that it is ready to use. It is then stored or pumped along a network of pipes. These pipes will sent it to all the places where it is needed.

▷ Gas leaks can be dangerous. At the gas treatment plant, a strong smell is added to gas, so that any leaks are quickly detected.

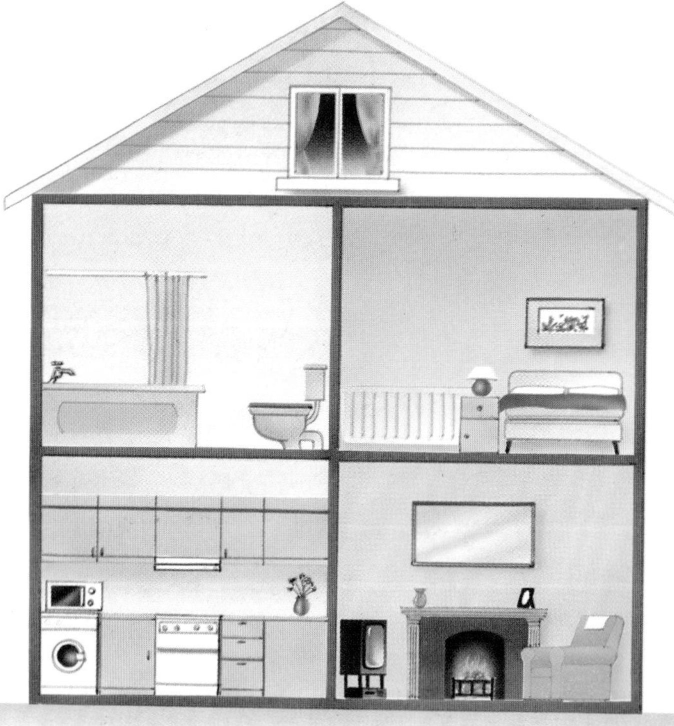

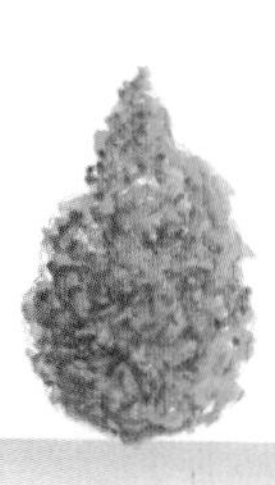

◁ Burning gas to make heat is a quick and easy way to warm the home and to cook. This house has gas central heating and a gas fireplace. There is a gas stove in the kitchen.

How Was Coal Formed?

Coal is found in layers, called seams, sandwiched between other layers of rock. About 350 million years ago, many parts of the Earth were covered with wet, swampy ground. Forests of trees and ferns grew in the swamps. When the plants died, they fell into the soft mud where they partly rotted and turned into **peat**. The peat became buried and slowly turned into coal.

▽ There are many different kinds of coal. The most useful coal is found deep underground.

▷ The trees and ferns sank into the swamp. Their remains were covered in mud. Over millions of years they slowly changed into peat.

▷ The mounting layers of rock pressed down harder and harder. The pressure and the heat changed the peat into coal.

Mining Coal

Geologists may find coal as deep as one thousand feet (305 m) under the ground. The mining company then cuts deep passages down to the coal seam. There have to be at least two passages. Then fresh air can move around. Tunnels are then bored from the bottom of the passages to the coal face. That is where parts of the coal seam are being cut.

▷ Most coal is deep underground, but a seam may be pushed closer to the surface by movements in the Earth's crust.

▽ The roof and sides of the tunnels are held up by strong supports.

H111
604
DEMAG

An Underground Factory

A coal mine is like an underground factory. There are workshops near the bottom of the shaft. Cars carry miners to the part of the coal face they are working on. That may be several miles from the shaft. Miners travel up and down the shaft in elevators. On the surface, there are offices, bathrooms, and a snack bar for the miners.

▽ Underground cars take miners to the coal face.

▷ A coal mine may not look like much from the outside, but underground there is a huge network of tunnels and a busy railroad system.

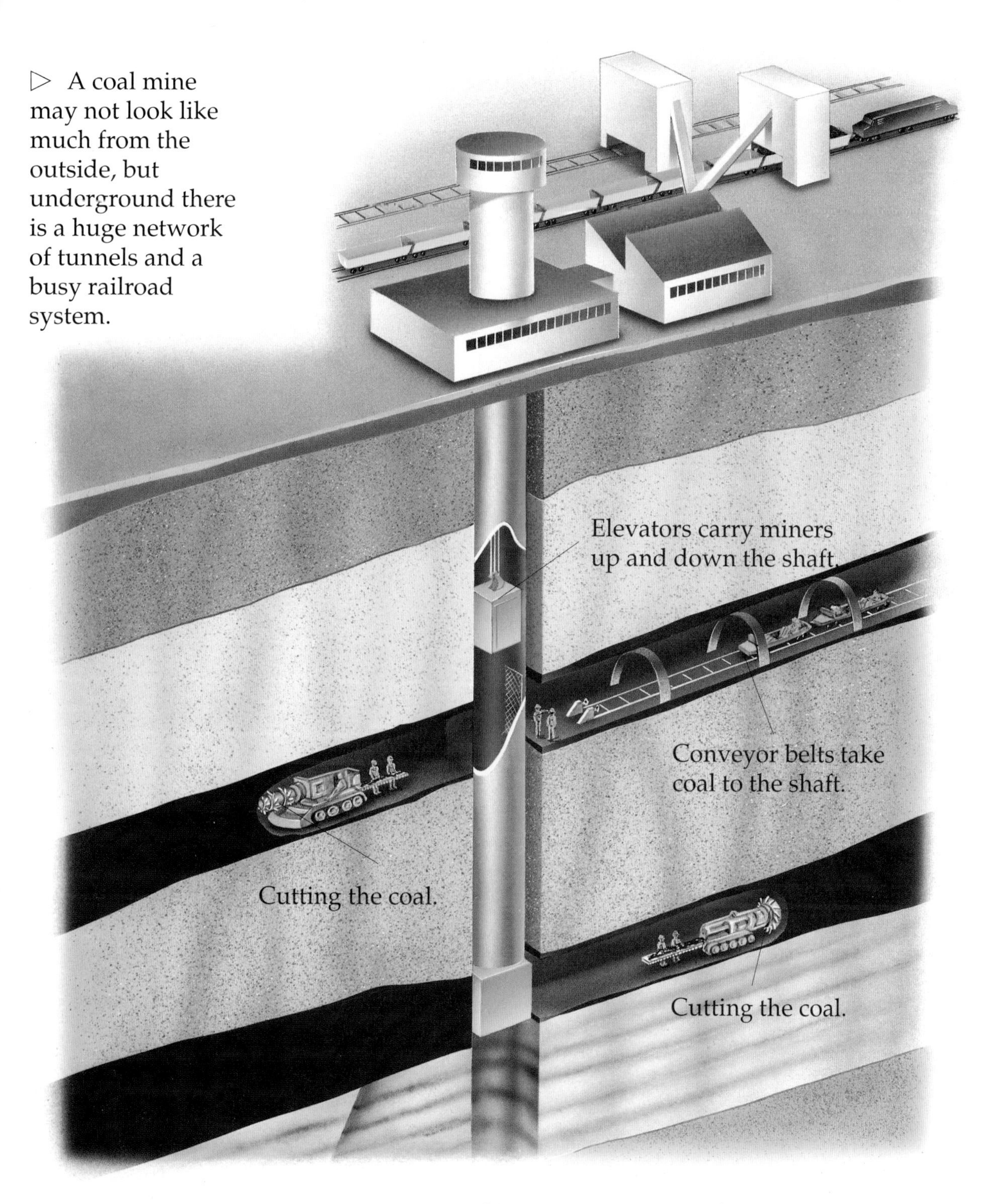
Elevators carry miners up and down the shaft.
Conveyor belts take coal to the shaft.
Cutting the coal.
Cutting the coal.

Cutting the Coal

Work at a mine goes on day and night. The miners work in shifts. Workers on the surface need to know who is underground. The miners are given tags that tell who they are. They hand these in when they finish work. For hundreds of years, miners had to cut coal by hand. Now there are drills and cutting machines to help them.

▷ A miner wears coveralls, boots, and a hard hat with a lamp on it. He has a mask to protect him from poisonous gases underground.

▽ A machine, like a massive drill, cuts tunnels through the rock.

▷ The coal cutter cuts chunks of coal in a long strip. The roof supports are moved forward to support the new tunnel.

Bringing the Coal Out

The cut coal is loaded onto a **conveyer belt** that takes it to a coal train. The train takes the coal to the bottom of the shaft where it is put into shuttle cars and lifted to the surface. There is a preparation plant nearby, where the coal is cleaned. It is then loaded onto trains or trucks and taken to other parts of the country or to a port for shipping overseas.

▷ Coal is loaded onto a bulk carrier ship for export to other countries. Coal is also carried by barge along rivers and canals.

▽ Deep in the mine, the coal is loaded onto a train that takes it to the bottom of the shaft. It is then brought up to the surface for cleaning.

KOCH

Surface Mining

Some coal is found near the surface of the Earth. Big earth-moving machines and explosives dig it out. One method, which is known as strip mining, is widely used in the United States and Australia. Surface mining is the cheapest kind of mining, but it can leave huge ugly scars in the landscape.

▽ A strip mine. The soil has to be removed to reach the coal. Rocks and soil are scraped or blasted away.

▷ Mining companies are now under pressure to restore the land they have mined.

▷ When the coal has been mined, the company returns the surface soil. The area is then landscaped and planted.

Using Coal

Many power plants use huge amounts of coal to produce electricity. In fact, they are often built near coal mines. Coal is moved from the mine to the power plant on conveyer belts or on trains. The train is loaded and unloaded automatically as it goes around. Coal is also used to make iron and steel, and as fuel for heating in people's homes.

▷ In huge steelmaking furnaces, part of the coal is used with other materials to make steel. The most widely used metal in the world is steel.

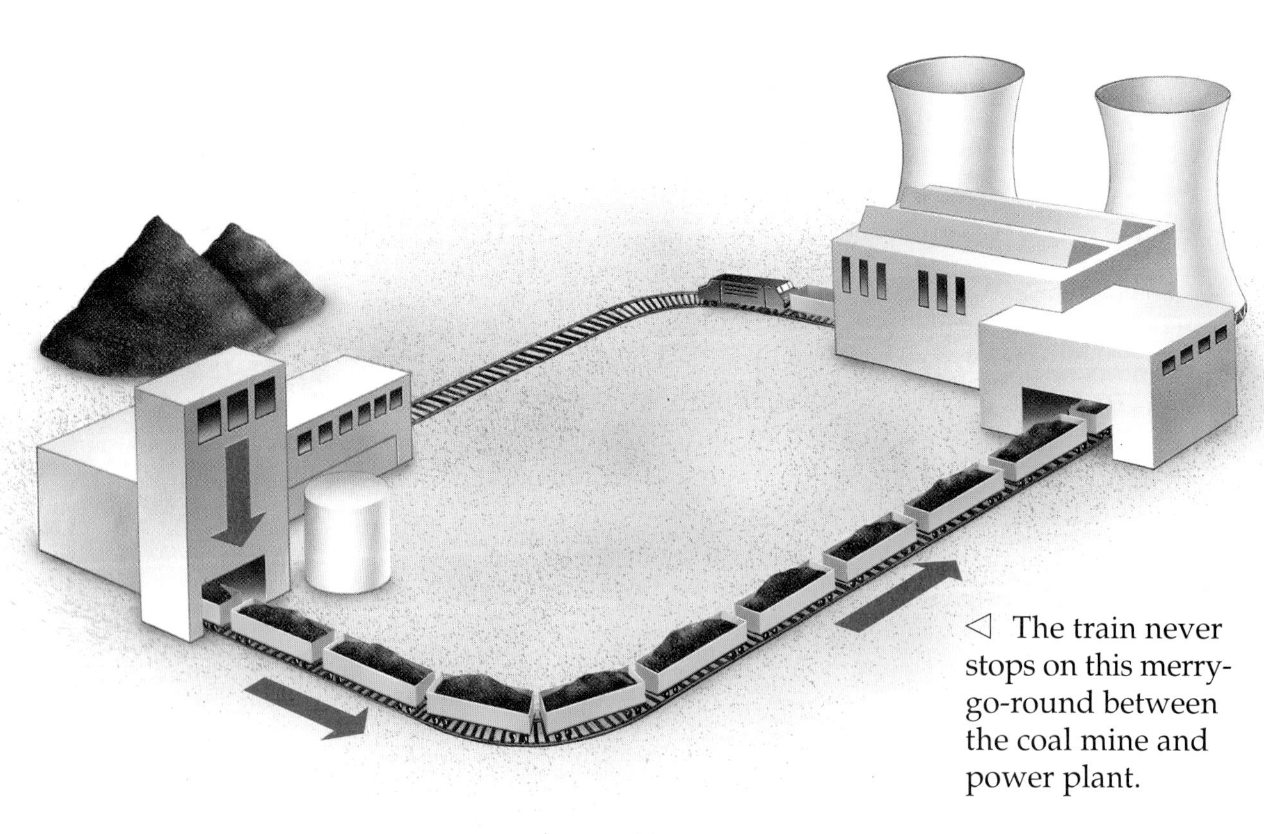

◁ The train never stops on this merry-go-round between the coal mine and power plant.

Things to Do

- Make a survey of all the things that use oil, gas, or coal in your home. Don't forget oil in machines and the family car. Which of these do you use most?

- Make a landscaping plan to cover a strip mine. Draw a picture of how you would use the land once the soil is put back. What kind of structures would you build?

Useful Addresses:

National Energy Information Center
U.S. Department of Energy
1000 Independence Avenue S.W.
Washington, DC 20585

American Petroleum Institute
1220 L Street N.W.
Washington, DC 20005

American Coal Foundation
1130 17th Street N.W., Suite 220
Washington, DC 20036-4604

American Gas Association
Educational Programs Department
1515 Wilson Boulevard
Arlington, VA 22209

Glossary

conveyer belt A belt that moves continuously to carry products from one place to another.

decayed Rotted away.

energy The ability to work or make things happen.

fossil The preserved remains of a dead animal or plant. Often fossils are found in rock.

fuel A material that can be burned to provide heat or light or to make machinery work.

geologist A scientist who studies the rocks on and under the surface of the Earth.

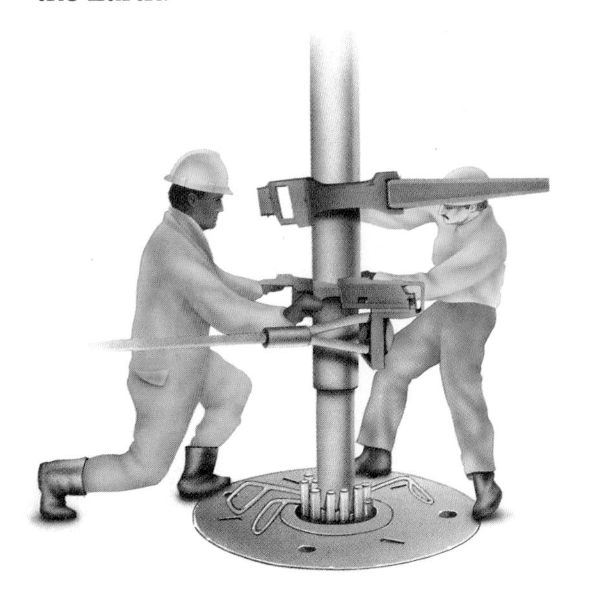

peat Layers of partly rotted dead plants that have been pressed tightly together. Peat is used as a fuel in some places.

pressure Pressing down or pushing up with a force or weight.

refinery A place where crude oil is made into oil products such as gasoline and diesel oil.

sedimentary rock Rock made from material that has settled to the bottom of a river, lake, or sea.

survey A study of the Earth's surface and the rock layers underneath.

Index

asphalt 13

coal 3, 16-29
  cleaning 24
  formation 3, 16-17
  transporting 24
  uses 28-29

coal face 18, 20
coal mines 18-27
crude oil 12

diesel oil 12
drills and cutting machines 8, 22

electricity 28
energy 3

fossil fuels 3
fuel 3, 12, 28

gas 3, 4-5, 6, 14-15
  finding gas 6, 14
  formation 3, 4-5
  transporting 14
  uses 14
gas leaks 14
gas treatment plants 14
geological survey 6

iron and steel 28

kerosene 12

oil 3, 4-13
  drilling for oil 4, 8-9, 10-11
  finding oil 6-7, 10-11
  formation 3, 4-5
  oil products 12-13
  transporting 8, 12
  under dry land 6-9
  under the ocean floor 10-11
oil platforms 10-11
oil rigs 4, 10-11
oil tankers (road or railroad) 12
oil tankers (ships) 8
oil wells 8

power plants 28

refineries 8, 12-13

sedimentary rock 4, 7
strip mining 26-27

Photographic credits:
British Coal Board 16, 20; J. Allan Cash Ltd. 3, 26; Eye Ubiquitous 29, © Mark Newham 6, © Bruce Adams 10; Robert Harding Picture Library 15, 19, 25, © Paolo Koch 23; Science Photo Library © Richard Folwell 4; Techno-Gaz 9; ZEFA 13